John Tarilanyo Afa

Energy Sources and Electrical Generation

GRIN Verlag

Bibliografische Information der Deutschen Nationalbibliothek:

Die Deutsche Bibliothek verzeichnet diese Publikation in der Deutschen Nationalbibliografie; detaillierte bibliografische Daten sind im Internet über http://dnb.d-nb.de/ abrufbar.

Imprint:

Druck und Bindung: Books on Demand GmbH, Norderstedt Germany
ISBN: 978-3-656-41465-0

This book at GRIN:

http://www.grin.com/en/e-book/213049/energy-sources-and-electrical-generation

JOHN TARILANYO AFA

UD12803SPO20130

CURRICULUM DESIGN
TECHNICAL REPORT

ESE 637:
ENERGY SOURCES AND ELECTRICAL GENERATION

ATLANTIC INTERNATIONAL UNIVERSITY

HONOLULU, HAWAII

2011

Atlantic International University

TABLE OF CONTENT

Atlantic International University

ENERGY SOURCES AND ELECTRICITY GENERATION IN NIGERIA

1.0 INTRODUCTION:

Electricity plays a very important role in the socio economic and technological development of every nation. The electricity demand in Nigeria far outstrips the supply and the supply also is epileptic in nature. The country is faced with acute electricity problems which are hindering its development, notwithstanding the availability of vast natural resources in the country.

The history of electricity in Nigeria dates back to 1896 where electricity was first produced in Lagos, fifteen years after its introduction in England. Despite the fact that its existence in the country is over a century, its development has been at a slow rate. In 1950, a central body was established by legislative council, which transferred electricity supply and development to the care of the central body known as the Electricity Corporation of Nigeria (ECN). Other bodies like the Native Authorities and Nigeria Electricity Supply Company (NESCO) have licenses to produces electricity in some locations in Nigeria.

There was another body known as Niger Dams Authority (NDA) established by an act of parliament. The Authority was responsible for the construction and maintenance of dams and other works on the River Niger and also generating electricity by means of water power, improving, production and promoting fish brims and irrigation. The energy produced by NDA was sold to the Electricity Corporation of Nigeria for distribution and sales at utility voltages.

In 1972 the National Electric Power Authority of Nigeria (NEPA) was established with the amalgamation of Electricity Corporation of Nigeria (ECN) and the Niger Dam Authority (NDA). NEPA was empowered to maintain an efficient coordinated and economically viable electricity supply in Nigeria (Sambo, 2008, Onwioduokit and Omoigui, 2000, ECN, 2004). In about 2005, the Power Holding of Nigeria gradually took over the functions of NEPA with little or no improvement to the power sector.

2.0 GENERAL ANALYSIS:

Within the particular conception of socio-economic processes which underscore every economic system, economic development globally revolves around the issues of the character, structure, pattern and evolution of desirable inter-personal relations of production, allocation and utilization of available resources in the country. In other to optimally develop and efficiently manage such available resources, equitably allocate and effectively utilize them and subsequently put economic development firmly on course, modern operational technologies with respect to production, allocation and utilization are designed and tied strictly to the use of energy in one form or the other. Thus, the quest to rapidly and firmly put the Nigerian economy on the course of economic development is technically, a function of adequate supply and distribution of energy, particularly electricity

In this regard, adequate supply and distribution of electricity constitute a central development issue which cannot be over-emphasized. Apart from serving as the pillar of wealth creation in Nigeria, it is also the nucleus of operations and subsequently the 'engine of growth' for all sectors of the economy. In recognition of the consolidating linkage between the energy sector and the other sectors of the economy, electricity development and utilization therefore have pervasive impacts on a range of socio-economic activities and consequently the living standard of citizens in the country.

From available data (Sambo, 2006 Nwankwo, 2000) about 40% only of the nation population of over 130 million have access to grid electricity. At the rural level, where 70 percent of the population lives, the availability of electricity is about 15 percent.

With the inadequate spread of power in the country, the National Electric Power Authority (NEPA) due to their inefficiency, electricity supply in Nigeria has been very unreliable so much that the demand on the so call available areas are not met. Indeed in most Nigerian towns hardly a day passes without an interruption in power supply and /or voltage and frequency variation. Sometimes, the outage last for hours. Some towns or part of towns experience complete blackout for days, sometimes months at a time. The incidence of Electricity outages, voltage and frequency variation in power system has become a rule rather than an exception in Nigeria. It has been observed that the problem centred on Nigeria's inability to co-ordinate the power resources resulting in lack of adequate generating capacity and too little expenditure on distribution facilities and maintenance. Other problems include inefficient load distribution, low capacity utilization, shortage of technical manpower in critical areas of the power industry and shortage or complete lack of essential spare parts as at when required. For over twenty years prior to 1999, the power sector did not witness substantial investment in infrastructural development. During that period, new plants were not constructed and existing ones were not properly maintained, bringing the power sector to a deplorable state. In 2001, generation went down from installed capacity of about 5600MW to an average of 1,750MW as compared to a load demand of 6,000MW. Also nineteen out of seventy nine (Sambo, 2008, Komolafe and Omoigui, 2000) installed generating units were in operation. The real installed units, installed capacity and the available units are shown in table 1:

Table 1: Installed Generating Capacity and units Available Units.

S/N	Power Station	Type	Installed Capacity	Units Installed	Units Available
1.	Egbin	Thermal	1320	6	3
2.	Sapele	Thermal	1020	10	2
3.	Delta	Thermal	912	20	4
4.	Kainji	Hydro	760	8	4
5.	Afarm	Thermal	636	17	1
6.	Shiroro	Hydro	600	4	4
7.	Jebba	Hydro	578.39	6	4
8.	Ikora and other	Thermal	74.19		
			5888.19	71	22

NEPA's installed generating capacity is made up of 1900MW hydro and 3,976MW thermal generation, Table 1 shows the list of generating stations and their availability. The national peak power demand to the connected people is about 3000MW that leaves a reserve of about 2,888MW which is about 49 percent of the installed capacity. The current available generating capacity is less than 2000MW (about 34%). This low available capacity is much less than the demand capacity, that is, 66 percent of the installed capacity is not available. Hence if all available units are put on line without any spinning reserve during the peak demand, the demand will be about 1000MW in excess of the available generation. Therefore load shedding is unavoidable.

Under construction: Due to the shortage of power supply, some state governments joined in building power station. The stations were built as federal projects. The on-going projects are given as shown in the table 2.

Table2: Power Stations under Construction.

Station	Type	Capacity MW	location
Genegu Power Station	Thermal	414	Kogi State
Omotosho Power Station	Thermal	335	Ondo State
Paplanto Power Station	Thermal	335	Ogun State
Alaoji Power Station	Thermal	378	Abia State
Omoku Power Station	Thermal	230	Rivers State
Gbarain/Ubie Power Station	Thermal	250	Bayelsa State
Sapele Power Station	Thermal	500	Delta State
Ikot Abasi Power Station	Thermal	270	Akwa Ibom State
Eyaen Power Station	Thermal	500	Edo State
Egbema Power Station	Thermal	250	Imo State
Calabar Power Station	Thermal	250	Cross Rivers State

3.0 THE INSTITUTIONAL SET UP AND THE POWER PROBLEM IN NIGERIA

NEPA is a partially commercialized monopoly operating in a non-contestable market. It is particular commercialized in that it is expected to fix tariff rates, charges and prices for its services to its consumers. The critical contradiction is the commercialization of what hitherto is a monopoly without either enacting an effective regulatory system to control the possible exploitation of the consumer or enacting an effective deregulatory system that allows the freedom of firm's entry and exit from the market. Between1983 to 1999, NEPA has adjusted its tariff five times (Nwankwo, 2000, Akarakiri, 2002)

It is necessary to mention that the partial commercialization of NEPA adopted by the government in 1999 has contributed to the problem of the Nigerian economy. This is because commercialization of a monopoly organization like NEPA can only translate into higher tariff without any improvement on services. To that end the industrialist pay more for the same level or in fact worsened level of unreliable power supply. According to the analysis of Onwioduokit, (Onwioduokit, 2000, Barr and Thomas 2005) the combined cost of unreliable NEPA and the cost of the personal owned generator increase the cost of production severally.

The Authority (NEPA) is a government parastatal that is responsible for generation, transmission and distribution of electricity in the country, therefore unlike a private sector do not have a clear profit objective and in the absence of profit motivation there is no incentive to minimize costs.

The Authority is always confident and faces no threat of bankruptcy as the government is always there to bail them out in event of financial crisis. One of the major reasons is the openness to manipulation by politicians who may set non commercial objectives for the enterprise in pursuit due to their political agenda.

Such political interferences can be extremely counterproductive and can lead to gross inefficiency.

Also, appointment to certain management cadre are subject to political associations, social connections, ethnicity or tribal sentiment that certain administrative in capabilities go unchecked, allow corruption and the organization become a money making venue for corrupt leaders and managers at the expense of the Nigerian populace.

The incentive structure confronting public sector management is often not compatible with the pursuit of productive efficiency since typically, neither their earnings nor tenure are directly linked to any measure of performance.

For political and social reasons public enterprises are seldom liquidated. Instead, they are cushioned by soft budget constraints and preferential access to both domestic and foreign credits. NEPA as one of the government parastatals suffer from over staffing without due consideration of need and technical manpower.

4.0 ENERGY RESOURCES IN NIGERIA

Nigeria is a primary energy store house accommodating such resources as coal and lignite, natural gas, crude oil, solar, hydro, nuclear, wood fuel, geothermal, tide biogas and Biomass. In spite of the vastness of these resources in Nigeria only natural gas and hydro are currently being utilized in process form while two other wood fuel and solar are used in their crude form for heating, cooking and lighting (MAN 2005, Fagbene et al, 2008)

4.1 Coal: Coal deposit was first discovered in the lowest Niger Basin of Nigeria in 1909 and it was first mined in 1916 when the Enugu coal fields were opened. From this discovery, the increasing local consumption of coal and the possibility of establishing markets in other West African countries led to an intensified search for coal in Nigeria.

As a result of these search several deposits of sub-bituminous coal belonging to the lower and upper ranks and lignite were discovered in different states of Nigeria. These include Anambra, Benue, Kwara, Bauchi, Plateau and Ondo. While the country is known to have over 1.5 billion coal reserves, only about 43 percent of these have been proven. Nigeria's coal is of high quality with low sulfur content.

4.2 Gas

Considerable reserves of natural gas have been discovered especially in the Niger Delta. The reserves are distributed between associated gas with oil and natural gas.

Nigeria currently has 256 trillion scf probable gas reserves, 120 trillion scf proven gas reserves and two billion associated gases produced daily. Out of these about 1.75 billion scf gas is flared every day (Akarakiri, 2002).

4.3 Crude Petroleum Product:

Crude Petroleum was first found in Nigeria in 1958 in Oloibiri and since then the number of oil reserves in the country has increased. Nigeria demonstrated crude oil reserves are currently well about 40 billion barrels.

4.4 Hydro-Electric Generation: There is an enormous potential for large, medium and small hydro-electric generation in the country. The paper presented by Nwankwo, J.O. (Nwankwo, 2000) the hydro-power development in Nigeria was given as shown in table 3.

Table 3: Potential Hydro-power Development in Nigeria

Site	River	Capacity MW
Lokoja-North	Niger	950
Onisha South	Niger	750
Makudi North	Benue	800
Zungeru I-North	Kaduna	500
Zungeru II-North	Kaduna	450
Yola-North	Benue	350
Kastina Ala-North	Kastina-Ala	260
Beli-North	Taraba	240
Garindala North	Taraba	135
Gembu-North	Donga	130
Ikom-North	Cross	400
Afikpo-South	Cross	180
Afan-South	Cross	180

Source: Electricity supply in Nigeria, Nwankwo 2000

The table 3 shows potential hydro power development from 1980 – 85. This study has been updated to include plans for other locations at Mambila North-hydro 1870MW, Onitsha South-hydro 1050MW, Ikerum South-hydro 737MW, Gurara North (Abuja) hydro 300MW, etc. There are series of studies going on but the underlining truth is that the potentials of hydro power is high.

4.5 **Solar:** Nigeria has an average yearly incident of solar energy on the ground of about 2300kwh / m^2. This gives total incident energy of about 2100 x 10^{12} kwh per year. The entire energy requirement of the country could be met by solar if one percent of the available solar energy can be tapped at 0.1 percent conversion efficiency [Olesen, 2004 Togola, 2005, Sambo, 2008).

Other renewable energies are wind, biomas, geothermal and ocean energies. The estimated capacities are given in table 4.

Table 4: Types of Energy Resources and Reserves

S/No	Resource Type		Reserves (Natural Units)		Reserves Energy Units (Btoe*)	Production	Domestic Utilization (Natural units)
1	Crude Oil		35 billion barrels		4.76	2.5 million barrels/day	450,000 barrels/day
2	Natural Gas		187 Trillion SCF		4.32	6 Billion SCF/day	3.4 billion SCF/day
3	Coal and lignite		2.175 billion tonnes		1.92	22.1 tonnes/day	22.1 tonnes/day
4	Tar Sands		31 billion barrels of equivalent		4.22	-	-
5	Hydropower large		15,000 MW		1.11 (over 38 years)	1,938 MW (167.4 million MWH/day	167.4 Million MWh/day
6	Small Hydropower		3,500 MW		0.25 (over 38 years)	30 MW (2.6 million MWh/day	2.6 million MWh/day
7	Solar Radiation		3.5 - 7.0 KWh/m²/day (485.1 million MWh/day using 0.1% Nigeria land area)		15.0 (38 years and 0.1% Nigeria land area)	Excess of 240 KWp of solar PV or 0.01 million MWh/day	Excess of 0.01million MWph/day of solar PV
8	Wind		(2-4) m/s at 10m height		8.14 (4m/s@ 70m height ⌀20m windmill, 0.1% land area of Nigeria over 38 years)	-	-
9	Biomass	Fuelwood	11 million hectares of forest and woodland	Excess of 1.2 tonnes/day	-	0.120 million tonnes/day	0.120 million tonnes/day
		Animal waste	211 million assorted animals		-	0.781 million tonnes of waste/day	Not available
		Energy Drops and Agric Residue	72 hectares of Agric. Land		-	0.256 million tonnes of assorted crops/day	Not available
10	Nuclear Element		Not yet qualified		-	-	-

Source: A.S. Sambo (2008)

5.0 DISCUSSION

From the analysis there are abundance of Energy Sources yet the country was going on with three hydro-station, Kainji, Jebba, and Shiroro; three gas turbine station Afam, Sapele and Delta; two steam turbine stations Sapele and Egbin.

Despite the abundant coal reserve in Nigeria there is not a single established coal fired power station. There are also several prime sites for hydro-electric power development in Nigeria which could be harnessed to produce relatively cheap power. For example, the proposed 950MW Zungeru Hydro-electric project downstream of Shiroro which has been fully engineered and awaiting funding since 1990 could be completed by the Independent Power Production. Due to government participation and the highly politicized nature of NEPA, backup with inconsistencies in the economy and political leadership of the Nation, the following points could be highlighted:

- Some management staff was appointed due to political and social affiliates, ethnicity or tribal sentiment.
- NEPA regards government to refurbish and rehabilitate the equipments and therefore make no effort of their own despite the high tariff collected.
- Expand their scope of operation therefore profit making is not their focus
- The high rates of tariff collected from consumers are misappropriated. It is unfortunate that NEPA depends on consumer to refurbish their transformer and service lines.
- Government companies, parastals and some government residential areas don't pay NEPA Bills.

- The commercial department of NEPA adjusts bills, deny customers the metering facilities to make gain for themselves.
- All these culminated to the inconsistency of NEPA, frustration to customer, create unemployment and deny of social amenities.

6.0 RECOMMENDATIONS

- The present government has strongly promised the privatization of the generation, transmission and the distribution levels. This is a welcome idea if reputable companies are involved especially at the generation and the distribution level and do not allow government official, Member of Parliament or those that feel they matter to be involved under any disguise.
- The billing system must be unified with good collection system of tariff.
- There must be an independent regulatory body to oversee the proper regulation of the guiding law concerning privatization. The interest of the consumer must be taken care off.
- Strict law concerning NEPA vandalisation and theft must be enacted and pursuit. In such cases proper education of the use of electric company facilities must be made and the legal implications must be addressed.
- The distribution systems must be reviewed and over hauled. This is important because most unqualified and greedy NEPA staff has made illegal connections that are detrimental to the system. The connections have made the system unbalanced because of these illegal connections.

7.0 CONCLUSION

For any stable power supply it is necessary to annest all available energy resources for maximum efficiency. The present energy system is from Hydro and Thermal. When there is a problem with water level the hydro-electrical generation will be affected. Also, the activities of militant in the Niger Delta affect the gas pipe line. It is therefore necessary to have an energy mix so as to guarantee stable power supply.

The energy commission of Nigeria with the energy policy for Nigeria in 2030 has planned gradual rise for all available energy sources in the country.

This endemic will be forgotten if this policy is followed properly. The questions that are in the mind of every Nigerian is how many times such millennium goals have been set yet little or no improvement has been made.

It is necessary to add that with this pragmatic approach of the present government the nation may have some relief but the long term approach if followed properly will give the nation the desired result in electricity production and consumption.

8.0 REFERENCES

1. Akarakiri, J.B (2002): Rural Energy in Nigeria. The Electricity Alternative: Domestic use of Energy Conference 2002: 1-7.
2. Barr., S., Thomas, JP. (2005). The role of renewable energy in the development of productive activities in rural West Africa: the case of Senegal.
3. ECN (2004): National Energy Policy, Federal Republic of Nigeria Published by the Energy Commission (ECN).
4. Fagbenle, R.Q, Adeleja A.O., Bellow, A.K, (2008). Modeling of Wired Energy Potential in Nigeria. Final Report of the Power Sector Reform committee, Abuja.
5. Hart, H.I. (2000). Gas Turbine Operation in a system disturbance – Prone Grid: 4th International Conference on Power Systems Operative and Planning Accra: 153 - 157
6. Jgbinovia, S.O. Orukpe, P.E., (2007). Rural Electrification: the propelling force for rural development of Edo State, Nigeria.
7. Komolafe, O.A., Omoigui, M.O., (2006). An Assessment of Reliability of Electricity Supply in Nigeria: 4th International Conference on Power Systems Operations and Planning Accra – Ghana.
8. Manufacturers Association of Nigeria (Man) (2005). Power Supply Reliability indices. A presentation at the Electric Power Sector reform workshop, Abuja.
9. Nwankwo, O.I., (2000). Electricity Supply in Nigeria: 4th International Conference on Power Systems OPration and Planning Accra, Ghana: 147 - 150
10. Olesen, G.B., (2004). Inforse Vision: 100% renewable in 2050, Sustainable Energy News, 45.
11. Onwioduokit, A. (2000). Privatization of Power Utility in Nigeria; Issues and options: 4th International Conference on Power Systems Operation and Planning, Accra, Ghana: 36 - 40
12. Sambo, A.S. (2006): Renewable Energy Electricity in Nigeria the way forward; Renewable Electricity Policy Conference:. 1 – 42.
13. Sambo, A.S (2008): Matching Electricity Supply with Demand in Nigeria; International Association for Energy Economics: 321 - 336
14. Sambo, A.S (2008): The Role of Energy in Achieving Millennium Development Goals (MDGs) National Engineering Tech. Conference (NETEC): 1 - 42
15. Togola, I., (2005): Renewable energy solution perspectives for Africa Mali – Folke Centre.